Deadly City

by

Paul W. Fairman

DEADLY CITY
by Paul W. Fairman

Copyright © 2023

ISBN: 978-93-59325-69-9

Published by

DOUBLE 9 BOOKS

2/13-B, Ansari Road, Daryaganj
New Delhi – 110002
info@double9books.com
www.double9books.com
Tel. 011-40042856

ABOUT THE AUTHOR

Paul Warren Fairman (1909-1977) was an editor and writer who worked under various pen names and under his real name. His "Late Rain" detective story appeared in the February 1947 edition of Mammoth Detective. His story "No Teeth for the Tiger" appeared in the February 1950 issue of Amazing Stories. He became the founding editor of If two years later, but only edited four issues. He took over as editor of Amazing Stories and Fantastic in 1955. He maintained both positions until 1958. The films "Deadly City" and "The Cosmic Frame" were based on his science fiction short stories. Fairman left Ziff Davis, the magazines' publisher, in 1958 when he was hired as managing editor of Ellery Queen's Mystery Magazine by its new publisher B. G. Davis, who had left ZD to found his own Davis Publications and purchased EQMM from Mercury Press as his first major act; Fairman continued until 1963, when he left to focus on writing his own work, often under different names. He ghost-wrote many children's novels, including The Runaway Robot (1965), which were based on outlines by Lester del Rey, whose name appeared on the publications. He also contributed to Ellery Queen's A Study in Terror (1966), in which Ellery discovers a previously undisclosed Sherlock Holmes manuscript.

You're all alone in a deserted city. You walk down an empty street, yearning for the sight of one living face – one moving figure. Then you see a man on a corner and you know your terror has only begun.

He awoke slowly, like a man plodding knee-deep through the thick stuff of nightmares. There was no definite line between the dream-state and wakefulness. Only a dawning knowledge that he was finally conscious and would have to do something about it.

He opened his eyes, but this made no difference. The blackness remained. The pain in his head brightened and he reached up and found the big lump they'd evidently put on his head for good measure – a margin of safety.

They must have been prudent people, because the bang on the head had hardly been necessary. The spiked drink which they had given him would have felled an ox. He remembered going down into the darkness after drinking it, and of knowing what it was. He remembered the helpless feeling.

It did not worry him now. He was a philosophical person, and the fact he was still alive cancelled out the drink and its result. He thought, with savor, of the chestnut-haired girl who had watched him take the drink. She had worn a very low bodice, and that was where his eyes had been at the last moment – on the beautiful, tanned breasts – until they'd wavered and puddled into a blur and then into nothing.

The chestnut-haired girl had been nice, but now she was gone and there were more pressing problems.

He sat up, his hands behind him at the ends of stiff arms clawing into long-undisturbed dust and filth. His movement stirred the dust and it rose into his nostrils.

He straightened and banged his head against a low ceiling. The pain made him sick for a minute and he sat down to regain his senses. He cursed the ceiling, as a matter of course, in an agonized whisper.

Ready to move again, he got onto his hands and knees and crawled cautiously forward, exploring as he went. His hand pushed through cobwebs and found a rough, cement wall. He went around and around. It was all cement – all solid.

Hell! They hadn't sealed him up in this place! There had been a way in so there had to be a way out. He went around again.

Then he tried the ceiling and found the opening—a wooden trap covering a four-by-four hole—covering it snugly. He pushed the trap away and daylight streamed in. He raised himself up until he was eye-level with a discarded shaving cream jar lying on the bricks of an alley. He could read the trade mark on the jar, and the slogan: "For the Meticulous Man".

He pulled himself up into the alley. As a result of an orderly childhood, he replaced the wooden trap and kicked the shaving cream jar against a garbage can. He rubbed his chin and looked up and down the alley.

It was high noon. An uncovered sun blazed down to tell him this.

And there was no one in sight.

He started walking toward the nearer mouth of the alley. He had been in that hole a long time, he decided. This conviction came from his hunger and the heavy growth of beard he'd sprouted. Twenty-four hours—maybe longer. That mickey must have been a lulu.

He walked out into the cross street. It was empty. No people—no cars parked at the curbs—only a cat washing its dirty face on a tenement stoop across the street. He looked up at the tenement windows. They stared back. There was an empty, deserted look about them.

The cat flowed down the front steps of the tenement and away toward the rear and he was truly alone. He rubbed his harsh chin. Must be Sunday, he thought. Then he knew it could not be Sunday. He'd gone into the tavern on a Tuesday night. That would make it five days. Too long.

He had been walking and now he was at an intersection where he could look up and down a new street. There were no cars—no people. Not even a cat.

A sign overhanging the sidewalk said: Restaurant. He went in under the sign and tried the door. It was locked. There were no lights inside. He turned away—grinning to reassure himself. Everything was all right. Just some kind of a holiday. In a big city like Chicago the people go away on hot summer holidays. They go to the beaches and the parks and sometimes you can't see a living soul on the streets. And of course you can't find any cars because the people use them to drive to the beaches and the parks and out into the country. He breathed a little easier and started walking again.

Sure—that was it. Now what the hell holiday was it? He tried to remember. He couldn't think of what holiday it could be. Maybe they'd dreamed up a new one. He grinned at that, but the grin was a little tight and he had to force it. He forced it carefully until his teeth showed white.

Pretty soon he would come to a section where everybody hadn't gone to the beaches and the parks and a restaurant would be open and he'd get a good meal.

A meal? He fumbled toward his pockets. He dug into them and found a handkerchief and a button from his cuff. He remembered that the button had hung loose so he'd pulled it off to keep from losing it. He hadn't lost the button, but everything else was gone. He scowled. The least they could have done was to leave a man eating money.

He turned another corner — into another street — and it was like the one before. No cars — no people — not even any cats.

Panic welled up. He stopped and whirled around to look behind him. No one was there. He walked in a tight circle, looking in all directions. Windows stared back at him — eyes that didn't care where everybody had gone or when they would come back. The windows could wait. The windows were not hungry. Their heads didn't ache. They weren't scared.

He began walking and his path veered outward from the sidewalk until he was in the exact center of the silent street. He walked down the worn white line. When he got to the next corner he noticed that the traffic signals were not working. Black, empty eyes.

His pace quickened. He walked faster — ever faster until he was trotting on the brittle pavement, his sharp steps echoing against the buildings. Faster. Another corner. And he was running, filled with panic, down the empty street.

The girl opened her eyes and stared at the ceiling. The ceiling was a blur but it began to clear as her mind cleared. The ceiling became a surface of dirty, cracked plaster and there was a feeling of dirt and squalor in her mind.

It was always like that at these times of awakening, but doubly bitter now, because she had never expected to awaken again. She reached down and pulled the wadded sheet from beneath her legs and spread it over them. She looked at the bottle on the shabby bed-table. There were three sleeping pills left in it. The girl's eyes clouded with resentment. You'd think seven pills would have done it. She reached down and took the sheet in both hands and drew it taut over her stomach. This was a gesture of frustration. Seven hadn't been enough, and here she was again — awake in the world she'd wanted to leave. Awake with the necessary edge of determination gone.

She pulled the sheet into a wad and threw it at the wall. She got up and walked to the window and looked out. Bright daylight. She wondered how long she had slept. A long time, no doubt.

Her naked thigh pressed against the windowsill and her bare stomach touched the dirty pane. Naked in the window, but it didn't matter, because it gave onto an airshaft and other windows so caked with grime as to be of no value as windows.

But even aside from that, it didn't matter. It didn't matter in the least.

She went to the washstand, her bare feet making no sound on the worn rug. She turned on the faucets, but no water came. No water, and she had a terrible thirst. She went to the door and had thrown the bolt before she remembered again that she was naked. She turned back and saw the half-empty Pepsi-Cola bottle on the floor beside the bed table. Someone else had left it there — how many nights ago? — but she drank it anyhow, and even though it was flat and warm it soothed her throat.

She bent over to pick up garments from the floor and dizziness came, forcing her to the edge of the bed. After a while it passed and she got her legs into one of the garments and pulled it on.

Taking cosmetics from her bag, she went again to the washstand and tried the taps. Still no water. She combed her hair, jerking the comb through the mats and gnarls with a satisfying viciousness. When the hair fell into its natural, blond curls, she applied powder and lip-stick. She went back to the bed, picked up her brassiere and began putting it on as she walked to the cracked, full-length mirror in the closet door. With the brassiere in place, she stood looking at her slim image. She assayed herself with complete impersonality.

She shouldn't look as good as she did — not after the beating she'd taken. Not after the long nights and the days and the years, even though the years did not add up to very many.

I could be someone's wife, she thought, with wry humor. I could be sending kids to school and going out to argue with the grocer about the tomatoes being too soft. I don't look bad at all.

She raised her eyes until they were staring into their own images in the glass and she spoke aloud in a low, wondering voice. She said, "Who the hell am I, anyway? Who am I? A body named Linda — that's who I am. No — that's *what* I am. A body's not a *who* — it's a *what*. One hundred and fourteen pounds of well-built blond body called Linda — model 1931 — no fender dents — nice paint job. Come in and drive me away. Price tag — "

She bit into the lower lip she'd just finished reddening and turned quickly to walk to the bed and wriggle into her dress — a gray and green cotton — the only one she had. She picked up her bag and went to the door. There she stopped to turn and thumb her nose at the three sleeping pills in the bottle before she went out and closed the door after herself.

The desk clerk was away from the cubbyhole from which he presided over the lobby, and there were no loungers to undress her as she walked toward the door.

Nor was there anyone out in the street. The girl looked north and south. No cars in sight either. No buses waddling up to the curb to spew out passengers.

The girl went five doors north and tried to enter a place called Tim's Hamburger House. As the lock held and the door refused to open, she saw that there were no lights on inside — no one behind the counter. The place was closed.

She walked on down the street followed only by the lonesome sound of her own clicking heels. All the stores were closed. All the lights were out.

All the people were gone.

He was a huge man, and the place of concealment of the Chicago Avenue police station was very small — merely an indentation low in the cement wall behind two steam pipes. The big man had lain in this niche for forty-eight hours. He had slugged a man over the turn of a card in a poolroom pinochle game, had been arrested in due course, and was awaiting the disposal of his case.

He was sorry he had slugged the man. He had not had any deep hatred for him, but rather a rage of the moment that demanded violence as its outlet. Although he did not consider it a matter of any great importance, he did not look forward to the six month's jail sentence he would doubtless be given.

His opportunity to hide in the niche had come as accidentally and as suddenly as his opportunity to slug his card partner. It had come after the prisoners had been advised of the crisis and were being herded into vans for transportation elsewhere. He had snatched the opportunity without giving any consideration whatever to the crisis. Probably because he did not have enough imagination to fear anything — however terrible — which might occur in the future. And because he treasured his freedom above all else. Freedom for today, tomorrow could take care of itself.

Now, after forty-eight hours, he writhed and twisted his huge body out of the niche and onto the floor of the furnace room. His legs were numb and he found that he could not stand. He managed to sit up and was able to bend his back enough so his great hands could reach his legs and begin to massage life back into them.

So elementally brutal was this man that he pounded his legs until they were black and blue, before feeling returned to them. In a few minutes he was walking out of the furnace room through a jail house which should now be utterly deserted. But was it? He went slowly, gliding along close to the walls to reach the front door unchallenged.

He walked out into the street. It was daylight and the street was completely deserted. The man took a deep breath and grinned. "I'll be damned," he muttered. "I'll be double and triple damned. They're all gone. Every damn one of them run off like rats and I'm the only one left. I'll be damned!"

A tremendous sense of exultation seized him. He clenched his fists and laughed loud, his laugh echoing up the street. He was happier than he had ever been in his quick, violent life. And his joy was that of a child locked in a pantry with a huge chocolate cake.

He rubbed a hand across his mouth, looked up the street, began walking. "I wonder if they took all the whisky with them," he said. Then he grinned; he was sure they had not.

He began walking in long strides toward Clark Street. In toward the still heart of the empty city.

He was a slim, pale-skinned little man, and very dangerous. He was also very clever. Eventually they would have found out, but he had been clever enough to deceive them and now they would never know. There was great wealth in his family, and with the rest of them occupied with leaving the city and taking what valuables they could on such short notice, he had been put in charge of one of the chauffeurs.

The chauffeur had been given the responsibility of getting the pale-skinned young man out of the city. But the young man had caused several delays until all the rest were gone. Then, meekly enough, he had accompanied the chauffeur to the garage. The chauffeur got behind the wheel of the last remaining car — a Cadillac sedan — and the young man had gotten into the rear seat.

But before the chauffeur could start the motor, the young man hit him on the head with a tire bar he had taken from a shelf as they had entered the garage.

The bar went deep into the chauffeur's skull with a solid sound, and thus the chauffeur found the death he was in the very act of fleeing.

The young man pulled the dead chauffeur from the car and laid him on the cement floor. He laid him down very carefully, so that he was in the exact center of a large square of outlined cement with his feet pointing straight north and his outstretched arms pointing south.

The young man placed the chauffeur's cap very carefully upon his chest, because neatness pleased him. Then he got into the car, started it, and headed east toward Lake Michigan and the downtown section.

After traveling three or four miles, he turned the car off the road and drove it into a telephone post. Then he walked until he came to some high weeds. He lay down in the weeds and waited.

He knew there would probably be a last vanguard of militia hunting for stragglers. If they saw a moving car they would investigate. They would take him into custody and force him to leave the city.

This, he felt, they had no right to do. All his life he had been ordered about—told to do this and that and the other thing. Stupid orders from stupid people. Idiots who went so far as to claim the whole city would be destroyed, just to make people do as they said. God! The ends to which stupid people would go in order to assert their wills over brilliant people.

The young man lay in the weeds and dozed off, his mind occupied with the pleasant memory of the tire iron settling into the skull of the chauffeur.

After a while he awoke and heard the cars of the last vanguard passing down the road. They stopped, inspected the Cadillac and found it serviceable. They took it with them, but they did not search the weeds along the road.

When they had disappeared toward the west, the young man came back to the road and began walking east, in toward the city.

Complete destruction in two days?

Preposterous.

The young man smiled.

The girl was afraid. For hours she had walked the streets of the empty city and the fear, strengthened by weariness, was now mounting toward terror. "One face," she whispered. "Just one person coming out of a house or walking across the street. That's all I ask. Somebody to tell me what this is all about. If I can find one person, I won't be afraid any more."

And the irony of it struck her. A few hours previously she had attempted suicide. Sick of herself and of all people, she had tried to end her own life. Therefore, by acknowledging death as the answer, she should now have no fear whatever of anything. Reconciled to crossing the bridge into death, no facet of life should have held terror for her.

But the empty city did hold terror. One face — one moving form was all she asked for.

Then, a second irony. When she saw the man at the corner of Washington and Wells, her terror increased. They saw each other at almost the same moment. Both stopped and stared. Fingers of panic ran up the girl's spine. The man raised a hand and the spell was broken. The girl turned and ran, and there was more terror in her than there had been before.

She knew how absurd this was, but still she ran blindly. What had she to fear? She knew all about men; all the things men could do they had already done to her. Murder was the ultimate, but she was fresh from a suicide attempt. Death should hold no terrors for her.

She thought of these things as the man's footsteps sounded behind her and she turned into a narrow alley seeking a hiding place. She found none and the man turned in after her.

She found a passageway, entered with the same blindness which had brought her into the alley. There was a steel door at the end and a brick lying by the sill. The door was locked. She picked up the brick and turned. The man skidded on the filthy alley surface as he turned into the areaway.

The girl raised the brick over her head. "Keep away! Stay away from me!"

"Wait a minute! Take it easy. I'm not going to hurt you!"

"Get away!"

Her arm moved downward. The man rushed in and caught her wrist. The brick went over his shoulder and the nails of her other hand raked his face. He seized her without regard for niceties and they went to the ground. She fought with everything she had and he methodically neutralized all her weapons — her hands, her legs, her teeth — until she could not move.

"Leave me alone. Please!"

"What's wrong with you? I'm not going to hurt you. But I'm not going to let you hit me with a brick, either!"

"What do you want? Why did you chase me?"

"Look — I'm a peaceful guy, but I'm not going to let you get away. I spent all afternoon looking for somebody. I found you and you ran away. I came after you."

"I haven't done anything to you."

"That's silly talk. Come on — grow up! I said I'm not going to hurt you."

"Let me up."

"So you can run away again? Not for a while. I want to talk to you."

"I—I won't run. I was scared. I don't know why. You're hurting me."

He got up—gingerly—and lifted her to her feet. He smiled, still holding both her hands. "I'm sorry. I guess it's natural for you to be scared. My name's Frank Brooks. I just want to find out what the hell happened to this town."

He let her withdraw her hands, but he still blocked her escape. She moved a pace backward and straightened her clothing. "I don't know what happened. I was looking for someone too."

He smiled again. "And then you ran."

"I don't know why. I guess—"

"What's your name."

"Nora—Nora Spade."

"You slept through it too?"

"Yes ... yes. I slept through it and came out and they were all gone."

"Let's get out of this alley." He preceded her out, but he waited for her when there was room for them to walk side by side, and she did not try to run away. That phase was evidently over.

"I got slipped a mickey in a tavern," Frank Brooks said. "Then they slugged me and put me in a hole."

His eyes questioned. She felt their demand and said, "I was—asleep in my hotel room."

"They overlooked you?"

"I guess so."

"Then you don't know anything about it?"

"Nothing. Something terrible must have happened."

"Let's go down this way," Frank said, and they moved toward Madison Street. He had taken her arm and she did not pull away. Rather, she walked invitingly close to him.

She said, "It's so spooky. So ... empty. I guess that's what scared me."

"It would scare anybody. There must have been an evacuation of some kind."

"Maybe the Russians are going to drop a bomb."

Frank shook his head. "That wouldn't explain it. I mean, the Russians wouldn't let us know ahead of time. Besides, the army would be here. Everybody wouldn't be gone."

"There's been a lot of talk about germ warfare. Do you suppose the water, maybe, has been poisoned?"

He shook his head. "The same thing holds true. Even if they moved the people out, the army would be here."

"I don't know. It just doesn't make sense."

"It happened, so it has to make sense. It was something that came up all of a sudden. They didn't have much more than twenty-four hours." He stopped suddenly and looked at her. "We've got to get out of here!"

Nora Spade smiled for the first time, but without humor. "How? I haven't seen one car. The buses aren't running."

His mind was elsewhere. They had started walking again. "Funny I didn't think of that before."

"Think of what?"

"That anybody left in this town is a dead pigeon. The only reason they'd clear out a city would be to get away from certain death. That would mean death is here for anybody that stays. Funny. I was so busy looking for somebody to talk to that I never thought of that."

"I did."

"Is that what you were scared of?"

"Not particularly. I'm not afraid to die. It was something else that scared me. The aloneness, I guess."

"We'd better start walking west — out of the city. Maybe we'll find a car or something."

"I don't think we'll find any cars."

He drew her to a halt and looked into her face. "You aren't afraid at all, are you?"

She thought for a moment. "No, I guess I'm not. Not of dying, that is. Dying is a normal thing. But I was afraid of the empty streets — nobody around. That was weird."

"It isn't weird now?"

"Not — not as much."

"I wonder how much time we've got?"

Nora shrugged. "I don't know, but I'm hungry."

"We can fix that. I broke into a restaurant a few blocks back and got myself a sandwich. I think there's still food around. They couldn't take it all with them."

They were on Madison Street and they turned east on the south side of the street. Nora said, "I wonder if there are any other people still here—like us?"

"I think there must be. Not very many, but a few. They would have had to clean four million people out overnight. It stands to reason they must have missed a few. Did you ever try to empty a sack of sugar? Really empty it? It's impossible. Some of the grains always stick to the sack."

A few minutes later the wisdom of this observation was proven when they came to a restaurant with the front window broken out and saw a man and a woman sitting at one of the tables.

He was a huge man with a shock of black hair and a mouth slightly open showing a set of incredibly white teeth. He waved an arm and shouted, "Come on in! Come on in for crissake and sit down! We got beer and roast beef and the beer's still cold. Come on in and meet Minna."

This was different, Nora thought. Not eerie. Not weird, like seeing a man standing on a deserted street corner with no one else around. This seemed normal, natural, and even the smashed window didn't detract too much from the naturalness.

They went inside. There were chairs at the table and they sat down. The big man did not get up. He waved a hand toward his companion and said, "This is Minna. Ain't she something? I found her sitting at an empty bar scared to death. We came to an understanding and I brought her along." He grinned at the woman and winked. "We came to a real understanding, didn't we, Minna?"

Minna was a completely colorless woman of perhaps thirty-five. Her skin was smooth and pale and she wore no makeup of any kind. Her hair was drawn straight back into a bun. The hair had no predominating color. It was somewhere between light brown and blond.

She smiled a little sadly, but the laugh did not cover her worn, tired look. It seemed more like a gesture of obedience than anything else. "Yes. We came to an understanding."

"I'm Jim Wilson," the big man boomed. "I was in the Chicago Avenue jug for slugging a guy in a card game. They kind of overlooked me when they cleaned the joint out." He winked again. "I kind of helped them overlook me. Then I found Minna." There was tremendous relish in his words.

Frank started introductions which Nora Spade cut in on. "Maybe you know what happened?" she asked.

Wilson shook his head. "I was in the jug and they didn't tell us. They just started cleaning out the joint. There was talk in the bullpen — invasion or something. Nobody knew for sure. Have some beer and meat."

Nora turned to the quiet Minna. "Did you hear anything?"

"Naw," Wilson said with a kind of affectionate contempt. "She don't know anything about it. She lived in some attic dump and was down with a sore throat. She took some pills or something and when she woke up they were gone."

"I went to work and —" Minna began, but Wilson cut her off.

"She swabs out some joints on Chicago Avenue for a living and that was how she happened to be sitting in that tavern. It's payday, and Minna was waiting for her dough!" He exploded into laughter and slapped the table with a huge hand. "Can you beat that? Waiting for her pay at a time like this."

Frank Brooks set down his beer bottle. The beer was cold and it tasted good. "Have you met anybody else? There must be some other people around."

"Uh-uh. Haven't met anybody but Minna." He turned his eyes on the woman again, then got to his feet. "Come on, Minna. You and I got to have a little conference. We got things to talk about." Grinning, he walked toward the rear of the restaurant. Minna got up more slowly. She followed him behind the counter and into the rear of the place.

Alone with Nora, Frank said, "You aren't eating. Want me to look for something else?"

"No — I'm not very hungry. I was just wondering —"

"Wondering about what?"

"When it will happen. When whatever is going to happen — you know what I mean."

"I'd rather know *what's* going to happen. I hate puzzles. It's hell to have to get killed and not know what killed you."

"We aren't being very sensible, are we?"

"How do you mean?"

"We should at least act normal."

"I don't get it."

Nora frowned in slight annoyance. "Normal people would be trying to reach safety. They wouldn't be sitting in a restaurant drinking beer. We should be trying to get away. Even if it does mean walking. Normal people would be trying to get away."

Frank stared at his bottle for a moment. "We should be scared stiff, shouldn't we?"

It was Nora's turn to ponder. "I'm not sure. Maybe not. I know I'm not fighting anything inside — fear, I mean. I just don't seem to care one way or another."

"I care," Frank replied. "I care. I don't want to die. But we're faced with a situation, and either way it's a gamble. We might be dead before I finish this bottle of beer. If that's true, why not sit here and be comfortable? Or we might have time to walk far enough to get out of range of whatever it is that chased everybody."

"Which way do you think it is?"

"I don't think we have time to get out of town. They cleaned it out too fast. We'd need at least four or five hours to get away. If we had that much time the army, or whoever did it, would still be around."

"Maybe they didn't know themselves when it's going to happen."

He made an impatient gesture. "What difference does it make? We're in a situation we didn't ask to get in. Our luck put us here and I'm damned if I'm going to kick a hole in the ceiling and yell for help."

Nora was going to reply, but at that moment Jim Wilson came striding out front. He wore his big grin and he carried another half-dozen bottles of beer. "Minna'll be out in a minute," he said. "Women are always slower than hell."

He dropped into a chair and snapped the cap off a beer bottle with his thumb. He held the bottle up and squinted through it, sighing gustily. "Man! I ain't never had it so good." He tilted the bottle in salute, and drank.

The sun was lowering in the west now, and when Minna reappeared it seemed that she materialized from the shadows, so quietly did she move. Jim Wilson opened another bottle and put it before her. "Here — have a drink, baby."

Obediently, she tilted the bottle and drank.

"What do you plan to do?" Frank asked.

"It'll be dark soon," Wilson said. "We ought to go out and try to scrounge some flashlights. I bet the power plants are dead. Probably aren't any flashlights either."

"Are you going to stay here?" Nora asked. "Here in the Loop?"

He seemed surprised. "Why not? A man'd be a fool to walk out on all this. All he wants to eat and drink. No goddam cops around. The life of Reilly and I should walk out?"

"Aren't you afraid of what's going to happen?"

"I don't give a good goddam what's going to happen. What the hell! Something's always going to happen."

"They didn't evacuate the city for nothing," Frank said.

"You mean we can all get killed?" Jim Wilson laughed. "Sure we can. We could have got killed last week too. We could of got batted in the can by a truck anytime we crossed the street." He emptied his bottle, threw it accurately at a mirror over the cash register. The crash was thunderous. "Trouble with you people, you're worry warts," he said with an expansive grin. "Let's go get us some flashlights so we can find our way to bed in one of those fancy hotels."

He got to his feet and Minna arose also, a little tired, a little apprehensive, but entirely submissive. Jim Wilson said, "Come on, baby. I sure won't want to lose *you*." He grinned at the others. "You guys coming?"

Frank's eyes met Nora's. He shrugged. "Why not?" he said. "Unless you want to start walking."

"I'm too tired," Nora said.

As they stepped out through the smashed window, both Nora and Frank half-expected to see other forms moving up and down Madison Street. But there was no one. Only the unreal desolation of the lonely pavement and the dark-windowed buildings.

"The biggest ghost town on earth," Frank muttered.

Nora's hand had slipped into Frank's. He squeezed it and neither of them seemed conscious of the contact.

"I wonder," Nora said. "Maybe this is only one of them. Maybe all the other big cities are evacuated too."

Jim Wilson and Minna were walking ahead. He turned. "If you two can't sleep without finding out what's up, it's plenty easy to do."

"You think we could find a battery radio in some store?" Frank asked.

"Hell no! They'll all be gone. But all you'd have to do is snoop around in some newspaper office. If you can read you can find out what happened."

It seemed strange to Frank that he had not thought of this. Then he realized he hadn't tried very hard to think of anything at all. He was surprised, also, at his lack of fear. He's gone through life pretty much taking things as they came — as big a sucker as the next man — making more than his quota of mistakes and blunders. Finding himself completely alone in a deserted city for the first time in his life, he had naturally fallen prey to sudden fright. But that had gradually passed, and now he was able to accept the new reality fairly passively. He wondered if that wasn't pretty much the way of all people. New situations brought a surge of whatever emotion fitted the picture. Then the emotion subsided and the new thing became the ordinary.

This, he decided, was the manner in which humanity survived. Humanity took things as they came. Pile on enough of anything and it becomes the ordinary.

Jim Wilson had picked up a garbage box and hurled it through the window of an electric shop. The glass came down with a crash that shuddered up the empty darkening street and grumbled off into silence. Jim Wilson went inside. "I'll see what I can find. You stay out here and watch for cops." His laughter echoed out as he disappeared.

Minna stood waiting silently, unmoving, and somehow she reminded Frank of a dumb animal; an unreasoning creature with no mind of her own, waiting for a signal from her master. Strangely, he resented this, but at the same time could find no reason for his resentment, except the feeling that no one should appear as much a slave as Minna.

Jim Wilson reappeared in the window. He motioned to Minna. "Come on in, baby. You and me's got to have a little conference." His exaggerated wink was barely perceptible in the gloom as Minna stepped over the low sill into the store. "Won't be long, folks," Wilson said in high good humor, and the two of them vanished into the darkness beyond.

Frank Brooks glanced at Nora, but her face was turned away. He cursed softy under his breath. He said, "Wait a minute," and went into the store through the huge, jagged opening.

Inside, he could barely make out the counters. The place was larger than it had appeared from the outside. Wilson and Minna were nowhere about.

Frank found the counter he was looking for and pawed out several flashlights. They were only empty tubes, but he found a case of batteries in a panel compartment against the wall.

"Who's there?"

"Me. I came in for some flashlights."

"Couldn't you wait?"

"It's getting dark."

"You don't have to be so damn impatient." Jim Wilson's voice was hostile and surly.

Frank stifled his quick anger. "We'll be outside," he said. He found Nora waiting where he'd left her. He loaded batteries into four flashlights before Jim Wilson and Minna reappeared.

Wilson's good humor was back. "How about the Morrison or the Sherman," he said. "Or do you want to get real ritzy and walk up to the Drake?"

"My feet hurt," Minna said. The woman spoke so rarely, Frank Brooks was startled by her words.

"Morrison's the closest," Jim Wilson said. "Let's go." He took Minna by the arm and swung off up the street. Frank and Nora fell in behind.

Nora shivered. Frank, holding her arm, asked, "Cold?"

"No. It's just all — unreal again."

"I see what you mean."

"I never expected to see the Loop dark. I can't get used to it."

A vagrant, whispering wind picked up a scrap of paper and whirled it along the street. It caught against Nora's ankle. She jerked perceptibly and kicked the scrap away. The wind caught it again and spiralled it away into the darkness.

"I want to tell you something," she said.

"Tell away."

"I told you before that I slept through the — the evacuation, or whatever it was. That wasn't exactly true. I did sleep through it, but it was my fault. I put myself to sleep."

"I don't get it."

"I tried to kill myself. Sleeping tablets. Seven of them. They weren't enough."

Frank said nothing while they paced off ten steps through the dark canyon that was Madison Street. Nora wondered if he had heard.

"I tried to commit suicide."

"Why?"

"I was tired of life, I guess."

"What do you want — sympathy?"

The sudden harshness in his voice brought her eyes around, but his face was a white blur.

"No — no, I don't think so."

"Well, you won't get it from me. Suicide is silly. You can have troubles and all that—everybody has them—but suicide—why did you try it?"

A high, thin whine—a wordless vibration of eloquence—needled out of the darkness into their ears. The shock was like a sudden shower of ice water dashed over their bodies. Nora's fingers dug into Frank's arm, but he did not feel the cutting nails. "We're—there's someone out there in the street!"

Twenty-five feet ahead of where Frank and Nora stood frozen there burst the booming voice of Jim Wilson. "What the hell was that?" And the shock was dispelled. The white circle from Wilson's flash bit out across the blackness to outline movement on the far side of the street. Then Frank Brook's light, and Nora's, went exploring.

"There's somebody over there," Wilson bellowed. "Hey, you! Show your face! Quit sneaking around!"

Frank's light swept an arc that clearly outlined the buildings across the street and then weakened as it swung westward. There was something or someone back there, but obscured by the dimness. He was swept by a sense of unreality again.

"Did you see them?"

Nora's light beam had dropped to her feet as though she feared to point it out into the darkness. "I thought I saw something."

Jim Wilson was swearing industriously. "There was a guy over there. He ducked around the corner. Some damn fool out scrounging. Wish I had a gun."

Frank and Nora moved ahead and the four stood in a group. "Put out your lights," Wilson said. "They make good targets if the jerk's got any weapons."

They stood in the darkness, Nora holding tightly to Frank's arm. Frank said, "That was the damndest noise I ever heard."

"Like a siren?" Frank thought Jim Wilson spoke hopefully, as though wanting somebody to agree with him.

"Not like any I ever heard. Not like a whistle, either. More of a moan."

"Let's get into that goddam hotel and—"

Jim Wilson's words were cut off by a new welling-up of the melancholy howling. It had a new pattern this time. It sounded from many places; not nearer, Frank thought, than Lake Street on the north, but spreading outward and backward and growing fainter until it died on the wind.

Nora was shivering, clinging to Frank without reserve.

Jim Wilson said, "I'll be damned if it doesn't sound like a signal of some kind."

"Maybe it's a language — a way of communication."

"But who the hell's communicating?"

"How would I know?"

"We best get to that hotel and bar a few doors. A man can't fight in the dark — and nothing to fight with."

They hurried up the street, but it was all different now. Gone was the illusion of being alone; gone the sense of solitude. Around them the ghost town had come suddenly alive. Sinister forces more frightening than the previous solitude had now to be reckoned with.

"Something's happened — something in the last few minutes," Nora whispered.

Frank leaned close as they crossed the street to the dark silent pile that was the Morrison hotel. "I think I know what you mean."

"It's as though there was no one around and then, suddenly, they came."

"I think they came and went away again."

"Did you actually *see* anyone when you flashed your light?"

"No — I can't say positively that I did. But I got the impression there were figures out there — at least dozens of them — and that they moved back away from the light. Always just on the edge of it."

"I'm scared, Frank."

"So am I."

"Do you think it could all be imagination?"

"Those moans? Maybe the first one — I've heard of people imagining sounds. But not the last ones. And besides, we all heard them."

Jim Wilson, utterly oblivious of any subtle emanations in the air, boomed out in satisfaction: "We don't have to bust the joint open. The revolving door works."

"Then maybe we ought to be careful," Frank said. "Maybe somebody else is around here."

"Could be. We'll find out."

"Why are we afraid?" Nora whispered.

"It's natural, isn't it?" Frank melted the beam of his light with that of Jim Wilson. The white finger pierced the darkness inside. Nothing moved.

"I don't see why it should be. If there are people in there they must be as scared as we are."

Nora was very close to him as they entered.

The lobby seemed deserted. The flashlight beams scanned the empty chairs and couches. The glass of the deserted cages threw back reflections.

"The keys are in there," Frank said. He vaulted the desk and scanned the numbers under the pigeon holes.

"We'd better stay down low," Jim Wilson said. "Damned if I'm going to climb to the penthouse."

"How about the fourth floor?"

"That's plenty high enough."

Frank came out with a handful of keys. "Odd numbers," he said. "Four in a row."

"Well I'll be damned," Jim Wilson muttered. But he said no more and they climbed the stairs in silence. They passed the quiet dining rooms and banquet halls, and by the time they reached the fourth floor the doors giving off the corridors had assumed a uniformity.

"Here they are." He handed a key to Wilson. "That's the end one." He said nothing as he gave Minna her key, but Wilson grunted, "For crissake!" in a disgusted voice, took Minna's key and threw it on the floor.

Frank and Nora watched as Wilson unlocked his door. Wilson turned. "Well, goodnight all. If you get goosed by any spooks, just yell."

Minna followed him without a word and the door closed.

Frank handed Nora her key. "Lock your door and you'll be safe. I'll check the room first." He unlocked the door and flashed his light inside. Nora was close behind him as he entered. He checked the bathroom. "Everything clear. Lock your door and you'll be safe."

"Frank."

"Yes?"

"I'm afraid to stay alone."

"You mean you want me to—"

"There are two beds here."

His reply was slow in coming. Nora didn't wait for it. Her voice rose to the edge of hysteria. "Quit being so damned righteous. Things have changed! Can't you realize that? What does it matter how or where we sleep? Does the world care? Will it make a damn bit of difference to the world whether I strip stark naked in front of you?" A sob choked in her throat. "Or would that outrage your morality."

He moved toward her, stopped six inches away. "It isn't that. For God's sake! I'm no saint. It's just that I thought you—"

"I'm plain scared, and I don't want to be alone. To me that's all that's important."

Her face was against his chest and his arms went around her. But her own hands were fists held together against him until he could feel her knuckles, hard, against his chest. She was crying.

"Sure," Frank said. "I'll stay with you. Now take it easy. Everything's going to be all right."

Nora sniffled without bothering to reach for her handkerchief. "Stop lying. You know it isn't going to be all right."

Frank was at somewhat of a loss. This flareup of Nora's was entirely unexpected. He eased toward the place the flashlight had shown the bed to be. Her legs hit its edge and she sat down.

"You—you want me to sleep in the other one?" he asked.

"Of course," Nora replied with marked bitterness. "I'm afraid you wouldn't be very comfortable in with me."

There was a time of silence. Frank took off his jacket, shirt and trousers. It was funny, he thought. He'd spent his money, been drugged, beaten and robbed as a result of one objective—to get into a room alone with a girl. And a girl not nearly as nice as Nora at that. Now, here he was alone with a real dream, and he was tongue-tied. It didn't make sense. He shrugged. Life was crazy sometimes.

He heard the rustle of garments and wondered how much Nora was taking off. Then he dropped his trousers, forgotten, to the floor. "Did you hear that?"

"Yes. It's that—"

Frank went to the window, raised the sash. The moaning sound came in louder, but it was from far distance. "I think that's out around Evanston."

Frank felt a warmth on his cheek and he realized Nora was by his side, leaning forward. He put an arm around her and they stood unmoving in complete silence. Although their ears were straining for the sound coming down from the north, Frank could not be oblivious of the warm flesh under his hand.

Nora's breathing was soft against his cheek. She said, "Listen to how it rises and falls. It's almost as though they were using it to talk with. The inflection changes."

"I think that's what it is. It's coming from a lot of different places. It stops in some places and starts in others."

"It's so—weird."

"Spooky," Frank said, "but in a way it makes me feel better."

"I don't see how it could." Nora pressed closer to him.

"It does though, because of what I was afraid of. I had it figured out that the city was going to blow up—that a bomb had been planted that they couldn't find, or something like that. Now, I'm pretty sure it's something else. I'm willing to bet we'll be alive in the morning."

Nora thought that over in silence. "If that's the way it is—if some kind of invaders are coming down from the north—isn't it stupid to stay here? Even if we are tired we ought to be trying to get away from them."

"I was thinking the same thing. I'll go and talk to Wilson."

They crossed the room together and he left her by the bed and went on to the door. Then he remembered he was in his shorts and went back and got his trousers. After he'd put them on, he wondered why he'd bothered. He opened the door.

Something warned him—some instinct—or possibly his natural fear and caution coincided with the presence of danger. He heard the footsteps on the carpeting down the hall—soft, but unmistakably footsteps. He called, "Wilson—Wilson—that you?"

The creature outside threw caution to the winds, Frank sensed rather than heard a body hurtling toward the door. A shrill, mad laughter raked his ears and the weight of a body hit the door.

Frank drew strength from pure panic as he threw his weight against the panel, but perhaps an inch or two from the latch the door wavered from opposing strength. Through the narrow opening he could feel the hoarse breath of exertion in his face. Insane giggles and curses sounded through the black stillness.

Frank had the wild conviction he was losing the battle, and added strength came from somewhere. He heaved and there was a scream and he knew he had at least one finger caught between the door and the jamb. He threw his weight against the door with frenzied effort and heard the squash of the finger. The voice kited up to a shriek of agony, like that of a wounded animal.

Even with his life at stake, and the life of Nora, Frank could not deliberately slice the man's fingers off. Even as he fought the urge, and called himself a fool, he allowed the door to give slightly inward. The hand was jerked to safety.

At that moment another door opened close by and Jim Wilson's voice boomed: "What the hell's going on out here?"

Simultaneous with this, racing footsteps receded down the hall and from the well of the stairway came a whining cry of pain.

"Jumping jees!" Wilson bellowed. "We got company. We ain't alone!"

"He tried to get into my room."

"You shouldn't have opened the door. Nora okay?"

"Yeah. She's all right."

"Tell her to stay in her room. And you do the same. We'd be crazy to go after that coot in the dark. He'll keep 'til morning."

Frank closed the door, double-locked it and went back to Nora's bed. He could hear a soft sobbing. He reached down and pulled back the covers and the sobbing came louder. Then he was down on the bed and she was in his arms.

She cried until the panic subsided, while he held her and said nothing. After a while she got control of herself. "Don't leave me, Frank," she begged. "Please don't leave me."

He stroked her shoulder. "I won't," he whispered.

They lay for a long time in utter silence, each seeking strength in the other's closeness. The silence was finally broken by Nora.

"Frank?"

"Yes."

"Do you want me?"

He did not answer.

"If you want me you can have me, Frank."

Frank said nothing.

"I told you today that I tried to commit suicide. Remember?"

"I remember."

"That was the truth. I did it because I was tired of everything. Because I've made a terrible mess of things. I didn't want to go on living."

He remained silent, holding her.

As she spoke again, her voice sharpened. "Can't you understand what I'm telling you? I'm no good! I'm just a bum! Other men have had me! Why shouldn't you? Why should you be cheated out of what other men have had?"

He remained silent. After a few moments, Nora said, "For God's sake, talk! Say something!"

"How do you feel about it now? Will you try again to kill yourself the next chance you get?"

"No—no, I don't think I'll ever try it again."

"Then things must look better."

"I don't know anything about that. I just don't want to do it now."

She did not urge him this time and he was slow in speaking. "It's kind of funny. It really is. Don't get the idea I've got morals. I haven't. I've had my share of women. I was working on one the night they slipped me the mickey—the night before I woke up to this tomb of a city. But now—tonight—it's kind of different. I feel like I want to protect you. Is that strange?"

"No," she said quietly. "I guess not."

They lay there silently, their thoughts going off into the blackness of the sepulchral night. After a long while, Nora's even breathing told him she was asleep. He got up quietly, covered her, and went to the other bed.

But before he slept, the weird wailings from out Evanston way came again—rose and fell in that strange conversational cadence—then died away into nothing.

Frank awoke to the first fingers of daylight. Nora still slept. He dressed and stood for some moments with his hand on the door knob. Then he threw the bolt and cautiously opened the door.

The hallway was deserted. At this point it came to him forcibly that he was not a brave man. All his life, he realized, he had avoided physical danger and had refused to recognize the true reason for so doing. He had classified himself as a man who dodged trouble through good sense; that the truly civilized person went out of his way to keep the peace.

He realized now that that attitude was merely salve for his ego. He faced the empty corridor and did not wish to proceed further. But stripped of the life-long alibi, he forced himself to walk through the doorway, close the door softly, and move toward the stairs.

He paused in front of the door behind which Jim Wilson and Minna were no doubt sleeping. He stared at it wistfully. It certainly would not be a mark of cowardice to get Jim Wilson up under circumstances such as these. In fact, he would be a fool not to do so.

Stubbornness forbade such a move, however. He walked softly toward the place where the hallway dead-ended and became a cross-corridor. He made the turn carefully, pressed against one wall. There was no one in sight. He got to the stairway and started down.

His muscles and nerves tightened with each step. When he reached the lobby he was ready to jump sky-high at the drop of a pin.

But no one dropped any pins, and he reached the modernistic glass doorway to the drugstore with only silence screaming in his ears. The door was unlocked. One hinge squeaked slightly as he pushed the door inward.

It was in the drugstore that Frank found signs of the fourth-floor intruder. An inside counter near the prescription department was red with blood. Bandages and first-aid supplies had been unboxed and thrown around with abandon. Here the man had no doubt administered to his smashed hand.

But where had he gone? Asleep, probably, in one of the rooms upstairs. Frank wished fervently for a weapon. Beyond doubt there was not a gun left in the Loop.

A gun was not the only weapon ever created, though, and Frank searched the store and found a line of pocket knives still in neat boxes near the perfume counter.

He picked four of the largest and found, also, a wooden-handled, lead-tipped bludgeon, used evidently for cracking ice.

Thus armed, he went out through the revolving door. He walked through streets that were like death under the climbing sun. Through streets and canyons of dead buildings upon which the new daylight had failed to shed life or diminish the terror of the night past.

At Dearborn he found the door to the Tribune Public Service Building locked. He used the ice breaker to smash a glass door panel. The crash of the glass on the cement was an explosion in the screaming silence. He went inside. Here the sense of desolation was complete; brought sharply to focus, probably, by the pigeon holes filled with letters behind the want-ad counter. Answers to a thousand and one queries, waiting patiently for someone to come after them.

Before going to the basement and the back files of the Chicago Tribune, Frank climbed to the second floor and found what he thought might be there—a row of teletype machines with a file-board hooked to the side of each machine.

Swiftly, he stripped the copy sheets off each board, made a bundle of them and went back downstairs. He covered the block back to the hotel at a dog-trot, filled with a sudden urge to get back to the fourth floor as soon as possible.

He stopped in the drugstore and filled his pockets with soap, a razor, shaving cream and face lotion. As an afterthought, he picked up a lavish cosmetic kit that retailed, according to the price tag, for thirty-eight dollars plus tax.

He let himself back into the room and closed the door softly. Nora rolled over, exposing a shoulder and one breast. The breast held his gaze for a full minute. Then a feeling of guilt swept him and he went into the bathroom and closed the door.

Luckily, a supply tank on the roof still contained water and Frank was able to shower and shave. Dressed again, he felt like a new man. But he regretted not hunting up a haberdashery shop and getting himself a clean shirt.

Nora had still not awakened when he came out of the bathroom. He went to the bed and stood looking down at her for some time. Then he touched her shoulder.

"Wake up. It's morning."

Nora stirred. Her eyes opened, but Frank got the impression she did not really awaken for several seconds. Her eyes went to his face, to the window, back to his face.

"What time is it?"

"I don't know. I think it's around eight o'clock."

Nora stretched both arms luxuriously. As she sat up, her slip fell back into place and Frank got the impression she hadn't even been aware of her partial nudity.

She stared up at him, clarity dawning in her eyes, "You're all cleaned up."

"I went downstairs and got some things."

"You went out—alone?"

"Why not. We can't stay in here all day. We've got to hit the road and get out of here. We've overshot our luck already."

"But that—that man in the hall last night! You shouldn't have taken a chance."

"I didn't bump into him. I found the place he fixed his hand, down in the drugstore."

Frank went to the table and came back with the cosmetic set. He put it in Nora's lap. "I brought this up for you."

Surprise and true pleasure were mixed in her expression. "That was very nice. I think I'd better get dressed."

Frank turned toward the window where he had left the bundle of teletype clips. "I've got a little reading to do."

As he sat down, he saw, from the corner of his eye, a flash of slim brown legs moving toward the bathroom. Just inside the door, Nora turned. "Are Jim Wilson and Minna up yet?"

"I don't think so."

Nora's eyes remained on him. "I think you were very brave to go downstairs alone. But it was a foolish thing to do. You should have waited for Jim Wilson."

"You're right about it being foolish. But I had to go."

"Why?"

"Because I'm not brave at all. Maybe that was the reason."

Nora left the bathroom door open about six inches and Frank heard the sound of the shower. He sat with the papers in his hand wondering about the water. When he had gone to the bathroom the thought had never occurred to him. It was natural that it should. Now he wondered about it. Why was it still running? After a while he considered the possibility of the supply tank on the roof.

Then he wondered about Nora. It was strange how he could think about her personally and impersonally at the same time. He remembered her words of the previous night. They made her—he shied from the term. What was the old cliche? A woman of easy virtue.

What made a woman of that type, he wondered. Was it something inherent in their makeup? That partially opened door was symbolic somehow. He was sure that many wives closed the bathroom door upon their husbands; did it without thinking, instinctively. He was sure Nora had left it partially open without thinking. Could a behavior pattern be traced from such an insignificant thing?

He wondered about his own attitude toward Nora. He had drawn away from what she'd offered him during the night. And yet from no sense of disgust. There was certainly far more about Nora to attract than to repel.

Morals, he realized dimly, were imposed—or at least functioned—for the protection of society. With society gone—vanished overnight—did the moral code still hold?

If and when they got back among masses of people, would his feelings toward Nora change? He thought not. He would marry her, he told himself firmly, as quick as he'd marry any other girl. He would not hold what she was against her. I guess I'm just fundamentally unmoral myself, he thought, and began reading the news clips.

There was a knock on the door accompanied by the booming voice of Jim Wilson. "You in there! Ready for breakfast?"

Frank got up and walked toward the door. As he did so, the door to the bathroom closed.

Jim Wilson wore a two-day growth of beard and it didn't seem to bother him at all. As he entered the room he rubbed his hands together in great gusto. "Well, where'll we eat, folks? Let's pick the classiest restaurant in town. Nothing but the best for Minna here."

He winked broadly as Minna, expressionless and silent, followed him in exactly as a shadow would have followed him and sat primly down in a straight-backed chair by the wall.

"We'd better start moving south," Frank said, "and not bother about breakfast."

"Getting scared?" Jim Wilson asked.

"You're damn right I'm scared—now. We're right in the middle of a big no-man's-land."

"I don't get you."

At that moment the bathroom door opened and Nora came out. Jim Wilson forgot about the question he'd asked. He let forth a loud whistle of appreciation. Then he turned his eyes on Frank and his thought was crystal clear. He was envying Frank the night just passed.

A sudden irritation welled up in Frank Brooks, a distinct feeling of disgust. "Let's start worrying about important things—our lives. Or don't you consider your life very important?"

Jim Wilson seemed puzzled. "What the hell's got into you? Didn't you sleep good?"

"I went down the block this morning and found some teletype machines. I've just been reading the reports."

"What about that guy that tried to get into your room last night?"

"I didn't see him. I didn't see anybody. But I know why the city's been cleaned out." Frank went back to the window and picked up the sheaf on clips he had gone through. Jim Wilson sat down on the edge of the bed, frowning. Nora followed Frank and perched on the edge of the chair he dropped into.

"The city going to blow up?" Wilson asked.

"No. We've been invaded by some form of alien life."

"Is that what the papers said?"

"It was the biggest and fastest mass evacuation ever attempted. I pieced the reports together. There was hell popping around here during the two days we — we waited it out."

"Where did they all go?" Nora asked.

"South. They've evacuated a forty-mile strip from the lake west. The first Terran defense line is set up in northern Indiana."

"What do you mean — Terra."

"It's a word that means Earth — this planet. The invaders came from some other planet, they think — at least from no place on Earth."

"That's the silliest damn thing I ever heard of," Wilson said.

"A lot of people probably thought the same thing," Frank replied. "Flying saucers were pretty common. Nobody thought they were anything and nobody paid much attention. Then they hit — three days ago — and wiped out every living soul in three little southern Michigan towns. From there they began spreading out. They —"

Each of them heard the sound at the same time. A faint rumble, increasing swiftly into high thunder. They moved as one to the window and saw four jet planes, in formation, moving across the sky from the south.

"There they come," Frank said. "The fight's started. Up to now the army has been trying to get set, I suppose."

Nora said, "Is there any way we can hail them? Let them know —"

Her words were cut off by the horror of what happened. As they watched, the plane skimmed low across the Loop. At a point, approximately over Lake Street, Frank estimated, the planes were annihilated. There was a flash of blue fire coming in like jagged lightning to form four balls of fire around the planes. The fire balls turned, almost instantly, into globes of white smoke that drifted lazily away.

And that was all. But the planes vanished completely.

"What happened?" Wilson muttered. "Where'd they go?"

"It was as if they hit a wall," Nora said, her voice hushed with awe.

"I think that *was* what happened," Frank said. "The invaders have some kind of a weapon that holds us helpless. Otherwise the army wouldn't have established this no-man's-land and pulled out. The reports said we have them surrounded on all sides with the help of the lake. We're trying to keep them isolated."

Jim Wilson snorted. "It looks like we've got them right where they want us."

"Anyhow, we're damn fools to stick around here. We'd better head south."

Wilson looked wistfully about the room. "I guess so, but it's a shame — walking away from all this."

Nora was staring out the window, a small frown on her face. "I wonder who they are and where they came from?"

"The teletype releases were pretty vague on that."

She turned quickly. "There's something peculiar about them. Something really strange."

"What do you mean?"

"Last night when we were walking up the street. It must have been these invaders we heard. They must have been across the street. But they didn't act like invaders. They seemed — well, scared. I got the feeling they ran from us in panic. And they haven't been back."

Wilson said, "They may not have been there at all. Probably our imaginations."

"I don't think so," Frank cut in. "They were there and then they were gone. I'm sure of it."

"Those wailing noises. They were certainly signalling to each other. Do you suppose that's the only language they have?" Nora walked over and offered the silent Minna a cigarette. Minna refused with a shake of her head.

"I wish we knew what they looked like," Frank said. "But let's not sit here talking. Let's get going."

Jim Wilson was scowling. There was a marked sullenness in his manner. "Not Minna and me. I've changed my mind. I'm sticking here."

Frank blinked in surprise. "Are you crazy? We've run our luck out already. Did you see what happened to those planes?"

"The hell with the planes. We've got it good here. This I like. I like it a lot. We'll stay."

"Okay," Frank replied hotly, "but talk for yourself. You're not making Minna stay!"

Wilson's eyes narrowed. "I'm not? Look, buster—how about minding your own goddam business?"

The vague feelings of disgust Frank had had now crystallized into words. "I won't let you get away with it! You think I'm blind? Hauling her into the back room every ten minutes! Don't you think I know why? You're nothing but a damn sex maniac! You've got her terrorized until she's afraid to open her mouth. She goes with us!"

Jim Wilson was on his feet. His face blazed with rage. The urge to kill was written in the crouch of his body and the twist of his mouth. "You goddam nosey little squirt. I'll—"

Wilson charged across the short, intervening distance. His arms went out in a clutching motion.

But Frank Brooks wasn't full of knockout drops this time, and with a clear head he was no pushover. Blinded with rage, Jim Wilson *was* a pushover. Frank stepped in between his outstretched arms and slugged him squarely on top of the head with the telephone. Wilson went down like a felled steer.

The scream came from Minna as she sprang across the room. She had turned from a colorless rag doll into a tigress. She hit Frank square in the belly with small fists at the end of stiff, outstretched arms. The full force of her charge was behind the fists, and Frank went backward over the bed.

Minna did not follow up her attack. She dropped to the floor beside Jim Wilson and took his huge head in her lap. "You killed him," she sobbed. "You—you murderer! You killed him! You had no right!"

Frank sat wide-eyed. "Minna! For God's sake! I was helping you. I did it for you!"

"Why don't you mind your business? I didn't ask you to protect me? I don't need any protection—not from Jim."

"You mean you didn't mind the way he's treated you—"

"You've killed him—killed him—" Minna raised her head slowly. She looked at Frank as though she saw him for the first time. "You're a fool" she said dully. "A big fool. What right have you got to meddle with other people's affairs? Are you God or something, to run people's lives?"

"Minna—I—"

It was as though he hadn't spoken. "Do you know what it's like to have nobody? All your life to go on and grow older without anybody? I didn't have no one and then Jim came along and wanted me."

Frank walked close to her and bent down. She reacted like a tiger. "Leave him alone! Leave him alone! You've done enough!"

Nonplused, Frank backed away.

"People with big noses — always sticking them in. That's you. Was that any of your business what he wanted of me? Did I complain?"

"I'm sorry, Minna. I didn't know."

"I'd rather go into back rooms with him than stay in front rooms without nobody."

She began to cry now. Wordlessly — soundlessly, rocking back and forth with the huge man's bloody head in her lap. "Anytime," she crooned. "Anytime I would—"

The body in her arms stirred. She looked down through her tears and saw the small black eyes open. They were slightly crossed, unfocused as they were by the force of the blow. They straightened and Jim mumbled, "What the hell — what the hell—"

Minna's time for talking seemed over. She smiled — a smile hardly perceptible, as though it was for herself alone. "You're all right," she said. "That's good. You're all right."

Jim pushed her roughly away and staggered to his feet. He stood swaying for a moment, his head turning; for all the world like a bull blinded and tormented. Then his eyes focused on Frank.

"You hit me with the goddam phone."

"Yeah — I hit you."

"I'm gonna kill you."

"Look — I made a mistake." Frank picked up the phone and backed against the wall. "I hit you, but you were coming at me. I made a mistake and I'm sorry."

"I'll smash your goddam skull."

"Maybe you will," Frank said grimly. "But you'll work for it. It won't come easy."

A new voice bit across the room. "Cut it out. I'll do the killing. That's what I like best. Everybody quiet down."

They turned and saw a slim, pale-skinned young man in the open doorway. The door had opened quietly and no one had heard it. Now the pale young man was standing in the room with a small, nickle-plated revolver in his right hand.

The left hand was close down at his side. It was swathed generously in white bandage.

The young man chuckled. "The last four people in the world were in a room," he said, "and there was a knock on the door."

His chuckle deepened to one of pure merriment. "Only there wasn't a knock. A man just walked in with a gun that made him boss."

No one moved. No one spoke. The man waited, then went on: "My name is Leroy Davis. I lived out west and I always had a keeper because they said I wasn't quite right. They wanted me to pull out with the rest of them, but I slugged my keeper and here I am."

"Put down the gun and we'll talk it over," Frank said. "We're all in this together."

"No, we aren't. I've got a gun, so that makes me top man. You're all in it together, but I'm not. I'm the boss, and which one of you tried to cut my hand off last night."

"You tried to break in here yelling and screaming like a madman. I held the door. What else could I do?"

"It's all right. I'm not mad. My type—we may be nuts, but we never hold a grudge. I can't remember much about last night. I found some whisky in a place down the street and whisky drives me nuts. I don't know what I'm doing when I drink whisky. They say once about five years ago I got drunk and killed a little kid, but I don't remember."

Nobody spoke.

"I got out of it. They got me out some way. High priced lawyers got me out. Cost my dad a pile."

Hysteria had been piling up inside of Nora. She had held it back, but now a little of it spurted out from between her set teeth. "Do something, somebody. *Isn't anybody going to do anything?*"

Leroy Davis blinked at her. "There's nothing they can do, honey," he said in a kindly voice. "I've got the gun. They'd be crazy to try anything."

Nora's laugh was like the rattle of dry peas. She sat down on the bed and looked up at the ceiling and laughed. "It's crazy. It's all so crazy! We're sitting here in a doomed city with some kind of alien invaders all around us and we don't know what they look like. They haven't hurt us at all. We don't even know what they look like. We don't worry a bit about them because we're too busy trying to kill each other."

Frank Brooks took Nora by the arm. "Stop it! Quit laughing like that!"

Nora shook him off. "Maybe we need someone to take us over. It's all pretty crazy!"

"Stop it."

Nora's eyes dulled down as she looked at Frank. She dropped her head and seemed a little ashamed of herself. "I'm sorry. I'll be quiet."

Jim Wilson had been standing by the wall looking first at the newcomer, then back at Frank Brooks. Wilson seemed confused as to who his true enemy really was. Finally he took a step toward Leroy Davis.

Frank Brooks stopped him with a motion, but kept his eyes on Davis. "Have you seen anybody else?"

Davis regarded Frank with long, careful consideration. His eyes were bright and birdlike. They reminded Frank of a squirrel's eyes. Davis said, "I bumped into an old man out on Halstead Street. He wanted to know where everybody had gone. He asked me, but I didn't know."

"What happened to the old man?" Nora asked. She asked the question as though dreading to do it; but as though some compulsion forced her to speak.

"I shot him," Davis said cheerfully. "It was a favor, really. Here was this old man staggering down the street with nothing but a lot of wasted years to show for his efforts. He was no good alive, and he didn't have the courage to die." Davis stopped and cocked his head brightly. "You know — I think that's what's been wrong with the world. Too many people without the guts to die, and a law against killing them."

It had now dawned upon Jim Wilson that they were faced by a maniac. His eyes met those of Frank Brooks and they were — on this point at least — in complete agreement. A working procedure sprang up, unworded, between them. Jim Wilson took a slow, casual step toward the homicidal maniac.

"You didn't see anyone else?" Frank asked.

Davis ignored the question. "Look at it this way," he said. "In the old days they had Texas long horns. Thin stringy cattle that gave up meat as tough as leather. Do we have cattle like that today? No. Because we bred out the weak line."

Frank said, "There are some cigarettes on that table if you want one."

Jim Wilson took another slow step toward Davis.

Davis said, "We bred with intelligence, with a thought to what a steer was for and we produced a walking chunk of meat as wide as it is long."

"Uh-huh," Frank said.

"Get the point? See what I'm driving at? Humans are more important than cattle, but can we make them breed intelligently? Oh, no! That interferes with damn silly human liberties. You can't tell a man he can only have two kids. It's his God-given right to have twelve when the damn moron can't support three. Get what I mean?"

"Sure — sure, I get it."

"You better think it over, mister—and tell that fat bastard to quit sneaking up on me or I'll blow his brains all over the carpet!"

If the situation hadn't been so grim it would have appeared ludicrous. Jim Wilson, feeling success almost in his grasp, was balanced on tiptoe for a lunge. He teetered, almost lost his balance and fell back against the wall.

"Take it easy," Frank said.

"I'll take it easy," Davis replied. "I'll kill every goddam one of you—" he pointed the gun at Jim Wilson "—starting with him."

"Now wait a minute," Frank said. "You're unreasonable. What right have you got to do that? What about the law of survival? You're standing there with a gun on us. You're going to kill us. Isn't it natural to try anything we can to save our own lives?"

A look of admiration brightened Davis' eyes. "Say! I like you. You're all right. You're logical. A man can talk to you. If there's anything I like it's talking to a logical man."

"Thanks."

"Too bad I'm going to have to kill you. We could sit down and have some nice long talks together."

"Why do you want to kill us?" Minna asked. She had not spoken before. In fact, she had spoken so seldom during the entire time they'd been together that her voice was a novelty to Frank. He was inclined to discount her tirade on the floor with Wilson's head in her lap. She had been a different person then. Now she had lapsed back into her old shell.

Davis regarded thoughtfully. "Must you have a reason?"

"You should have a reason to kill people."

Davis said, "All right, if it will make you any happier. I told you about killing my keeper when they tried to make me leave town. He got in the car, behind the wheel. I got into the back seat and split his skull with a tire iron."

"What's that got to do with us?"

"Just this. Tommy was a better person than anyone of you or all of you put together. If he had to die, what right have you got to live? Is that enough of a reason for you?"

"This is all too damn crazy," Jim Wilson roared. He was on the point of leaping at Davis and his gun.

At that moment, from the north, came a sudden crescendo of the weird invader wailings. It was louder than it had previously been but did not seem nearer.

The group froze, all ears trained upon the sound. "They're talking again," Nora whispered.

"Uh-huh," Frank replied. "But it's different this time. As if—"

"—as if they were getting ready for something," Nora said. "Do you suppose they're going to move south?"

Davis said, "I'm not going to kill you here. We're going down stairs."

The pivotal moment, hinged in Jim Wilson's mind, that could have changed the situation, had come and gone. The fine edge of additional madness that would make a man hurl himself at a loaded gun, was dulled. Leroy Davis motioned pre-emptorily toward Minna.

"You first—then the other babe. You walk side by side down the hall with the men behind you. Straight down to the lobby."

They complied without resistance. There was only Jim Wilson's scowl, Frank Brooks' clouded eyes, and the white, taut look of Nora.

Nora's mind was not on the gun. It was filled with thoughts of the pale maniac who held it. He was in command. Instinctively, she felt that maniacs in command have one of but two motivations—sex and murder. Her reaction to possible murder was secondary. But what if this man insisted upon laying his hands upon her. What if he forced her into the age old thing she had done so often? Nora shuddered. But it was also in her mind to question, and be surprised at the reason for her revulsion. She visualized the hands upon her body—the old familiar things, and the taste in her mouth was one of horror.

She had never experienced such shrinkings before. Why now. Had she herself changed? Had something happened during the night that made the past a time of shame? Or was it the madman himself? She did not know.

Nora returned from her musings to find herself standing in the empty lobby. Leroy Davis, speaking to Frank, was saying, "You look kind of tricky to me. Put your hands on your head. Lock your fingers together over your head and keep your hands there."

Jim Wilson was standing close to the mute Minna. She had followed all the orders without any show of anger, with no outward expression. Always she had kept her eyes on Jim Wilson. Obviously, whatever Jim ordered, she would have done without question.

Wilson leaned his head down toward her. He said, "Listen, baby, there's something I keep meaning to ask but I always forget it. What's your last name?"

"Trumble—Minna Trumble. I thought I told you."

"Maybe you did. Maybe I didn't get it."

Nora felt the hysteria welling again. "How long are you going to keep doing this?" she asked.

Leroy Davis cocked his head as he looked at her. "Doing what?"

"Play cat and mouse like this. Holding us on a pin like flies in an exhibit."

Leroy Davis smiled brightly. "Like a butterfly in your case, honey. A big, beautiful butterfly."

"What are you going to do," Frank Brooks snapped. "Whatever it is, let's get it over with?"

"Can't you see what I'm doing?" Davis asked with genuine wonder. "Are you that stupid? I'm being the boss. I'm in command and I like it. I hold life and death over four people and I'm savoring the thrill of it. You're pretty stupid, mister, and if you use that 'can't get away with it' line, I'll put a bullet into your left ear and watch it come out your right one."

Jim Wilson's fists were doubled. He was again approaching the reckless point. And again it was dulled by the gradually increasing sound of a motor — not in the air, but from the street level to the south.

It was a sane, cheerful sound and was resented instantly by the insane mind of Leroy Davis.

He tightened even to the point that his face grew more pale from the tension. He backed to a window, looked out quickly, and turned back. "It's a jeep," he said. "They're going by the hotel. If anybody makes a move, or yells, they'll find four bodies in here and me gone. That's what I'm telling you and you know I'll do it."

They knew he *would* do it and they stood silent, trying to dredge up the nerve to make a move. The jeep's motor backfired a couple of times as it approached Madison Street. Each time, Leroy Davis' nerves reacted sharply and the four people kept their eyes trained on the gun in his hand.

The jeep came to the intersection and slowed down. There was a conference between its two occupants — helmeted soldiers in dark brown battle dress. Then the jeep moved on up Clark Street toward Lake.

A choked sigh escaped from Nora's throat. Frank Brooks turned toward her. "Take it easy," he said. "We're not dead yet. I don't think he wants to kill us."

The reply came from Minna. She spoke quietly. "I don't care. I can't stand any more of this. After all, we aren't animals. We're human beings and we have a right to live and die as we please."

Minna walked toward Leroy Davis. "I'm not afraid of your gun any more. All you can do with it is kill me. Go ahead and do it."

Minna walked up to Leroy Davis. He gaped at her and said, "You're crazy! Get back there. You're a crazy dame!"

He fired the gun twice and Minna died appreciating the incongruity of his words. She went out on a note of laughter and as she fell, Jim Wilson, with an echoing animal roar, lunged at Leroy Davis. His great hand closed completely over that of Davis, hiding the gun. There was a muffled explosion and the bullet cut unnoticed through Wilson's palm. Wilson jerked the gun from Davis' weak grasp and hurled it away. Then he killed Davis.

He did it slowly, a surprising thing for Wilson. He lifted Davis by his neck and held him with his feet off the floor. He squeezed Davis' neck, seeming to do it with great leisure as Davis made horrible noises and kicked his legs.

Nora turned her eyes away, buried them in Frank Brooks' shoulder, but she could not keep the sounds from reaching her ears. Frank held her close. "Take it easy," he said. "Take it easy." And he was probably not conscious of saying it.

"Tell him to hurry," Nora whispered. "Tell him to get it over with. It's like killing—killing an animal."

"That's what he is—an animal." Frank Brooks stared in fascination at Leroy Davis' distorted, darkening face. It was beyond semblance of anything human now. The eyes bulged and the tongue came from his mouth as though frantically seeking relief.

The animal sounds quieted and died away. Nora heard the sound of the body falling to the floor—a limp, soft sound of finality. She turned and saw Jim Wilson with his hands still extended and cupped. The terrible hands from which the stench of a terrible life was drifting away into empty air.

Wilson looked down at his handiwork. "He's dead," Wilson said slowly. He turned to face Frank and Nora. There was a great disappointment in his face. "That's all there is to it," he said, dully. "He's just—dead." Without knowing it for what it was, Jim Wilson was full of the futile aftertaste of revenge.

He bent down to pick up Minna's body. There was a small blue hole in the right cheek and another one over the left eye. With a glance at Frank and Nora, Jim Wilson covered the wounds with his hand as though they were not decent. He picked her up in his arms and walked across the lobby and up the stairs with the slow, quiet tread of a weary man.

The sound of the jeep welled up again, but it was further away now. Frank Brooks took Nora's hand and they hurried out into the street. As they crossed the sidewalk, the sound of the jeep was drowned by a sudden swelling of the wailings to the northward.

On still a new note, they rose and fell on the still air. A note of panic, of new knowledge, it seemed, but Frank and Nora were not paying close attention. The sounds of the jeep motor had come from the west and they got within sight of the Madison-Well intersection in time to see the jeep hurtle southward at its maximum speed.

Frank yelled and waved his arms, but he knew he had been neither seen nor heard. They were given little time for disappointment however, because a new center of interest appeared to the northward. From around the corner of Washington Street, into Clark, moved three strange figures.

There was a mixture of belligerence and distress in their actions. They carried odd looking weapons and seemed interested in using them upon something or someone, but they apparently lacked the energy to raise them although they appeared to be rather light.

The creatures themselves were humanoid, Frank thought. He tightened his grip on Nora's hand. "They've seen us."

"Let's not run," Nora said. "I'm tired of running. All it's gotten us is trouble. Let's just stand here."

"Don't be foolish."

"I'm not running. You can if you want to."

Frank turned his attention back to the three strange creatures. He allowed natural curiosity full reign. Thoughts of flight vanished from his mind.

"They're so thin — so fragile," Nora said.

"But their weapons aren't."

"It's hard to believe, even seeing them, that they're from another planet."

"How so? They certainly don't look much like us."

"I mean with the talk, for so long, about flying saucers and space flight and things like that. Here they are, but it doesn't seem possible."

"There's something wrong with them."

This was true. Two of the strange beings had fallen to the sidewalk. The third came doggedly on, dragging one foot after the other until he went to his hands and knees. He remained motionless for a long time, his head hanging limply. Then he too, sank to the cement and lay still.

The wailings from the north now took on a tone of intense agony — great desperation. After that came a yawning silence.

"They defeated themselves," the military man said. "Or rather, natural forces defeated them. We certainly had little to do with it."

Nora, Frank, and Jim Wilson stood at the curb beside a motorcycle. The man on the cycle supported it with a leg propped against the curb as he talked.

"We saw three of them die up the street," Frank said.

"Our scouting party saw the same thing happen. That's why we moved in. It's about over now. We'll know a lot more about them and where they came from in twenty-four hours."

They had nothing further to say. The military man regarded them thoughtfully. "I don't know about you three. If you ignored the evacuation through no fault of your own and can prove it—"

"There were four of us," Jim Wilson said. "Then we met another man. He's inside on the floor. I killed him."

"Murder?" the military man said sharply.

"He killed a woman who was with us," Frank said. "He was a maniac. When he's identified I'm pretty sure he'll have a past record."

"Where is the woman's body?"

"On a bed upstairs," Wilson said.

"I'll have to hold all of you. Martial law exists in this area. You're in the hands of the army."

The streets were full of people now, going about their business, pushing and jostling, eating in the restaurants, making electricity for the lights, generating power for the telephones.

Nora, Frank, and Jim Wilson sat in a restaurant on Clark Street. "We're all different people now," Nora said. "No one could go through what we've been through and be the same."

Jim Wilson took her statement listlessly. "Did they find out what it was about our atmosphere that killed them?"

"They're still working on that, I think." Frank Brooks stirred his coffee, raised a spoonful and let it drip back into the cup.

"I'm going up to the Chicago Avenue police station," Wilson said.

Frank and Nora looked up in surprise. Frank asked, "Why? The military court missed it—the fact you escaped from jail."

"They didn't miss it I don't think. I don't think they cared much. I'm going back anyway."

"It won't be much of a rap."

"No, a pretty small one. I want to get it over with."

He got up from his chair. "So long. Maybe I'll see you around."

"So long."

"Goodbye."

Frank said, "I think I'll beat it too. I've got a job in a factory up north. Maybe they're operating again." He got to his feet and stood awkwardly by the table. "Besides—I've got some pay coming."

Nora didn't say anything.

Frank said, "Well—so long. Maybe I'll see you around."

"Maybe. Goodbye."

Frank Brooks walked north on Clark Street. He was glad to get away from the restaurant. Nora was a good kid but hell—you didn't take up with a hooker. A guy played around, but you didn't stick with them.

But it made a guy think. He was past the kid stage. It was time for him to find a girl and settle down. A guy didn't want to knock around all his life.

Nora walked west on Madison Street. Then she remembered the Halstead Street slums were in that direction and turned south on Wells. She had nine dollars in her bag and that worried her. You couldn't get along on nine dollars in Chicago very long.

There was a tavern on Jackson near Wells. Nora went inside. The barkeep didn't frown at her. That was good. She went to the bar and ordered a beer and was served.

After a while a man came in. A middle aged man who might have just come into Chicago—whose bags might still be at the LaSalle Street Station down the block. The man looked at Nora, then away. After a while looked at her again.

Nora smiled.